WILDLIFE IN BLOOM SERIES

Little Opossum

BY AUTHOR & CONSERVATIONIST

LINDA BLACKMOOR

ISBN: 979-8-9904465-9-5 (PRINT)

PUBLISHED BY QUILL PRESS. LINDA BLACKMOOR'S TITLES MAY BE
PURCHASED IN BULK FOR EDUCATIONAL, BUSINESS, FUNDRAISING, OR
SALES PROMOTIONAL USE. FOR INFORMATION, PLEASE EMAIL
HELLO@LINDABLACKMOOR.COM

FIRST PRINT EDITION: 2024

LINDA BLACKMOOR
WWW.LINDABLACKMOOR.COM

MARSUPIAL

Opossums are the only marsupials native to North America, carrying their young in pouches like kangaroos. The Virginia opossum thrives from Canada to Central America, adapting to forests, fields, and urban areas. Leading a nomadic lifestyle, they rarely stay in one den for more than a few days, constantly exploring. Their diet of fruits, insects, and small animals allows them to thrive in diverse environments.

POUCH

Baby opossums, called joeys, are born the size of a honeybee and crawl into their mother's pouch to continue developing. A mother can carry up to 13 joeys, who cling to her for weeks before riding on her back. The pouch keeps them safe and warm while they grow and learn to explore the world. Only a few survive to adulthood, as life in the wild is full of challenges.

TAIL

Opossums use their long, prehensile tails like a fifth limb, gripping branches as they climb. While they don't hang upside down as cartoons suggest, their tails help them balance and carry nesting materials to their dens. Young opossums practice climbing using their tails, gaining strength and confidence. This remarkable adaptation is unique among North American mammals.

IMMUNITY

Opossums are nearly immune to snake venom, including rattlesnakes and cottonmouths. Their low body temperature makes them resistant to many diseases like rabies. They consume vast numbers of ticks —up to 5,000 in a season— helping control tick populations and reduce the spread of Lyme disease. Scientists study their unique biology to develop treatments for snakebites.

DEFENSE

When threatened, opossums "play dead" by entering a catatonic state that can last for hours, fooling predators. Their body goes limp, tongue hangs out, and they emit a foul-smelling fluid to mimic decay. This involuntary response protects them from predators like coyotes, foxes, and dogs. This theatrical defense is unique among mammals and showcases nature's creativity.

TEETH

Opossums have 50 teeth—the most of any North American land mammal—which they use to eat a variety of foods. Their sharp teeth are suited for crunching insects, small animals, and tough vegetation. Their omnivorous diet includes fruits, grains, and carrion, making them efficient scavengers. These dental adaptations allow them to thrive where food sources vary greatly.

NIGHT

With large, dark eyes, opossums are adapted for their nocturnal lifestyle, seeing well in dim light. Their vision helps them navigate forests, forage under moonlight, and avoid predators. Unlike some animals, their eyes lack a reflective layer, so they don't glow in headlights. Their sense of touch and hearing complement their eyesight, heightening their awareness.

METABOLISM

Opossums have a low body temperature, averaging 94–97°F, one of the lowest among mammals. This trait helps them resist many diseases but makes them vulnerable to frostbite in colder climates. To survive winter, they rely on dens, sheltered spaces, and thick fur coats, and may enter a state of torpor to conserve energy. Their unique metabolism ensures survival even in harsh conditions.

SURVIVORS

Opossums have existed for over 70 million years, coexisting with dinosaurs before adapting to modern ecosystems. Their lineage is one of the oldest among mammals, making them living fossils today. Despite ancient roots, they remain highly adaptable, thriving in urban, rural, and wild habitats. Their survival skills include climbing, scavenging, and clever escape tactics.

CLEAN

Opossums are meticulous groomers, spending hours cleaning their fur and removing parasites. This behavior helps them stay healthy and free from pests, contributing to their role as natural tick controllers. They avoid fouling their dens, ensuring a clean environment for themselves and their young. Their surprising cleanliness is a little-known trait of these misunderstood creatures.

SMELL

Opossums have a highly developed sense of smell, guiding them to food and danger in the night. Their noses detect faint scents, helping them find ripe fruit, hidden insects, and carrion. This keen sense aids them in identifying threats and navigating unfamiliar territory. Combined with excellent hearing, it makes them formidable foragers in the darkness.

SOLITARY

Opossums are solitary wanderers, spending much of their lives exploring alone rather than forming groups. Males and females come together briefly to mate, and mothers part ways with their young once they are independent. This solitary nature helps them avoid competition for resources and live in harmony with other wildlife. Their independence is key to their survival strategy.

ADAPT

Opossums thrive in urban areas, scavenging for food in gardens, garbage cans, and compost piles. Their ability to adapt to human landscapes demonstrates their resourcefulness and resilience. By consuming pests and scraps, they perform an essential cleanup service in cities. Though often misunderstood, they are gentle neighbors with a knack for coexistence.